About the Author:

I am the owner of YouTube maths channel Achieve Maths. My website is www.achievemaths.co.uk. I have experience of teaching Maths at all levels from ages 11 to 18 and have worked in schools and a sixth form college. I have been an expert examiner for GCSE and A-level Mathematics. I have experience of face-to-face and online tuition with my tutees achieving great success. I have experience outside of education in the financial sector and the NHS.

I truly believe that everyone has the potential to achieve in maths. I love the subject and I am on a mission to raise its profile and to help people to start understanding maths more.

If you have any queries or questions that you would like to ask me, either while reading or after reading my book, then feel free to email me on achievemaths@gmail.com. I have some, but limited, availability for private online tuition through Skype.

How to Achieve in Maths

Achieve Maths

The dedication of this book is to you, the reader. In purchasing this book, you have made a commitment, an endeavour to achieve where others do not. I expect more from you than just reading the pages. I encourage you to engage in the content and allow this book to change you.

Contents:

Introduction 1

I. A Common Current Understanding of Maths 2

II. My Vision of what Maths is and what it should be 6

III. How to Build a Mind-set that will help you to Achieve in Maths 10

IV. Action Steps to Achieve in Maths 13

V. Maths Questions that will make you think 18

VI. Model Solutions to Questions 28

Introduction:

I believe that mathematics is a subject that should be accessible to all. I believe that with the right mind-set and direction that everyone can be successful at mathematics to some degree. The aim of this book is to help you in this quest and to make sure that you achieve in maths, whatever that may mean for you.

As a lover of maths I am extremely disappointed in the general reputation that it has. Many people see it in a negative light and become disheartened by the very mention of the word maths, or are scared by the sight of numbers. I think that the reason for this is that the majority of people do not understand maths. I don't mean by this that they struggle to solve quadratic equations (although they might). I mean that they do not understand what the subject is about, what it is trying to do, why it is useful and how it should be approached.

In this book, I aim to explain what I think maths is all about and outline how you can change your attitude towards it. My perception of the subject may be more of a challenge but I hope that you will embrace the challenge and realise that even though it is difficult, it is something that you can work at and improve; to reach whatever level you choose. After all, nothing worth having is easy.

I. A Common Current Understanding of Maths

It is my perception that maths is seen as something that you are shown how to do and then you do. For example; I show you how to solve x + 7 = 10 by subtracting 7 from both sides of the equation and realising that x=3, then I ask you to solve x + 6 = 10 with the same method. Traditionally you would follow the exact same steps that were demonstrated and find that x=4. There is an additional problem with this particular example; many students would see that effectively all we are doing is thinking about "something" add 6 giving us 10 and would know that this something (x) must be 4.

This in essence is fine and it is good that a student has managed to find a different approach to solve this particular problem. However, with this example and question the teacher is not trying to teach you how to add up two numbers and get to 10. They are trying to teach you some of the fundamentals of algebra, and in particular the fundamentals of solving equations. In mathematical terms, they are trying to demonstrate that; you solve equations by applying the inverse of each operation (in the appropriate order) until you are left with only the unknown on one side of the equation. I know that this last sentence probably sounded very complicated and hard to understand, but that sentence is one thing that you need to be able to do to solve equations. This complication is why teachers of maths split this one concept into many examples of different cases.

If all you do as a student is learn all of the different cases individually then you may never understand what is really going on when you carry out the methods, and why they work. It is no wonder why people find maths boring, it can be very boring to do things over and over again without knowing what it all means. On the other hand, it can be satisfying to carry out algorithms and arrive at a correct result.

You might be able to solve some of the questions involving a given technique, but as soon as a question comes along that requires you to have understood what you are doing, then you will struggle. When I teach, I try to present the underlying understanding of a question, as well as demonstrating how to do it.

Many students want hard and fast ways to get the correct answer without any concern as to how or why. I think that part of the reason for this is that we live in an age where more importance is given to the result than to the process and everyone seems to want to progress quickly at any cost. In the past, it has been possible to get good grades at GCSE mathematics by effectively copying techniques that you have learnt / been taught. I have taught many students that can compute questions correctly, but become stuck when I ask them a question to demonstrate their actual understanding of the basic mathematical concept underpinning what they have done. I believe this is where things have to change because it is missing the vital trait of mathematics, which is to be used as a tool to deconstruct and solve problems.

Fortunately, it seems that exam boards and the government agree with my thinking, and hence exams are becoming "more difficult", i.e. they are going to require candidates to have in depth understanding of topics. They are going to require students to think for themselves and be able to attempt problems that they have never been shown how to do! I'm sorry to those of you that like to be shown how to do something and then copying it, but this is only scratching the surface of what it means to be good at maths.

Being good at maths means that you can take what you have been shown and apply it to somewhere that you have not been shown. An example of this type of application is if I were to tell you that the interior angles of a triangle add up to $180°$, then ask you what you think the interior angles of a four-sided shape (quadrilateral) would add up to. This question requires you to make connections between the information that you have been given and the answer required.

Some of you will already "know" the answer to this question and it would be straight forward to say that you know that the interior angles of a four-sided shape sum to $360°$, because you have been told this before. But I don't believe that by recalling a fact you are actively participating in mathematics. A more mathematical approach would be to realise that we can cut any quadrilateral into two triangles. This means that the interior angles of a four-sided shape are the same as the interior angles of the two triangles that go together to form it. By this logic, we arrive at our answer by adding the interior angles of the two triangles ($180° + 180° = 360°$). See figure 1.

Clearly, we get the same answer by remembering what we have been told or "figuring it out" for ourselves, but I believe that the latter is much more powerful. By training your brain to solve problems in this way you will not only become a better mathematician; it will help you in problem solving in your day to day life. If you are thinking to yourself, "wait a minute, how do we know the angles of a triangle add up to 180 degrees?" well, now you are beginning to think like a mathematician.

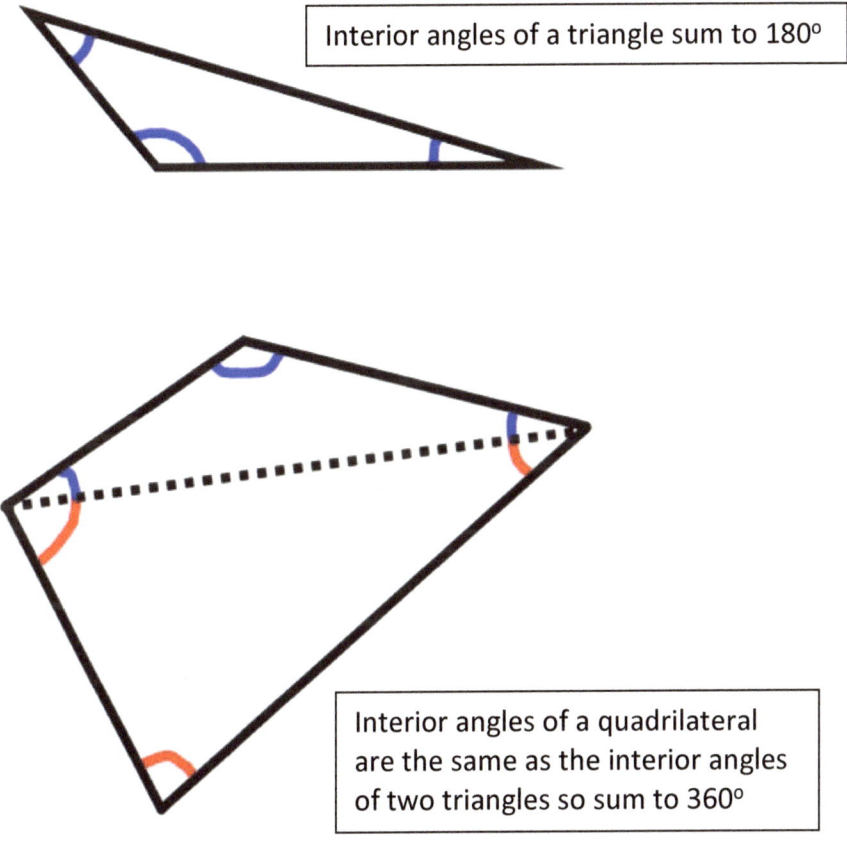

Interior angles of a triangle sum to 180°

Interior angles of a quadrilateral are the same as the interior angles of two triangles so sum to 360°

Figure 1. Combining Interior Angles

II. My Vision of what Maths is and what it should be

To me, maths is like a difficult computer game. If you realise that it is going to be difficult but you are willing to put in the time and effort to succeed, then you will get better and better. However, if you see how difficult the computer game is, maybe even see people playing it and doing really well, it is easy to become demoralised and give up. When I see people that are very good at computer games; I understand that they have not always been that good. To become good, they will have had to struggle, and work through difficulties and challenges. When I look at people who are good at maths, I see the same thing. At some point, they had to realise how to add up two numbers to make 10. At some point, they had to get their head around using letters to represent unknown or variable numbers. Maybe at some point, they had to learn about what complex numbers are and how to deal with them. My point is; everyone has to go down a similar path to become good at maths, but some people get a little stuck along the way. If you get stuck with understanding how we can use a letter in equations, then how can you hope to really understand how to solve equations!

I think maths is all about perseverance and problem solving; someone gives you a question that you do not already know the answer to. Then you have to use your brain power and any skills that you have learnt, to find the answer (or in some cases; an answer that you think is acceptable). This is what separates learning maths from learning other subjects. Learning that "hola" means "hello" in Spanish, or that the capital city of France is Paris, is learning facts. In contrast, good Maths teaching should teach skills, as well as hard facts. There are facts that need to be learnt in maths, and for your Maths exams you will need to learn certain formulae. But for real maths, it is far more important to understand and know how to use a formula, than it is to be able to remember it.

In real life, we have the internet to look up mathematical facts and formulae, but as mathematicians we need to know what they mean and how to use them. Let me walk you through a made-up example problem that you could come across in real life, and how I would use my mathematical skills to solve it. Don't worry too much if you cannot fully follow the maths, I just want to show you how a structured approach can be built up. I might be thinking about depositing £100 at the end of each month into a bank account which pays 1% interest every month. My maths skills would allow me to quickly analyse how much money I would have in my account after a certain number of months. I know that 1% is the same as 0.01 as a decimal (1/100) and that this decimal can be used as a multiplier to find 1% of a number. Since we want to add 1% to what we already have (100%), multiplying by 1.01 (100% + 1%) will automatically add 1% interest to the amount that we multiply by.

I might want to calculate, for example, how much money I should have in the account after three years (36 months). The first investment of £100 would get 35 months' interest, the next 34, the next 33 etc., down to the final £100 which would not get any. If we were to write these amounts out backwards as a series, we would get: £100 + £100 x 1.01 + £100 x 1.01^2 + ... + £100 x 1.01^{35}. Being able to add up this series is a skill needed for A-level maths, and the topic is called geometric series. However, the mathematical skill is to realise that the original problem can be written in this form, and that the geometric series formula can be used. I have used this example because right now I would have to go to Google and search for "sum of geometric series formula" to remind myself of the formula: $S_N = a(1 - r^N) /$ $(1 - r)$. Where N is the number of terms, S_N is the sum of those N terms, a is the first term in the series, and r is the common ratio (what we multiply by each time).

The answer is then 100 x $(1 - 1.01^{36})$ / (1 − 1.01) = £4,307.69. This is one reason why I love maths; we can solve almost any problem by breaking it into smaller and smaller parts, and by using formulae based on mathematical truths.

I know exactly where the formula for the sum of a geometric series comes from and how to use it. I can show a proof of why it works from first principles. This means that I am completely confident that it will get me the right answer, without having to check the whole sum on my calculator. Maths may appear mystical and magical, but it is actually very structured and methodical, there is a reason for everything. Because I have many things to remember, the actual formula may well slip out of my head, but I am not worried about this because I understand the formula. On the other hand; an A-level Maths student might have learnt the formula off by heart, but not understand how and when to use it fully. For day to day life, my approach is fine, however, for the artificial scenario of examinations you must learn the required formulae and fully understand them. If I was writing an exam, I would give all the required formulae and ask questions based on understanding, rather than memory. The internet is available almost anywhere you go, this acts as an extension to your own memory. It allows you to save room in your own head for more important things. The internet cannot think for you though (not yet anyway), so these skills that you develop will be important for the rest of your life.

III. How to Build a Mind-set that will help you to Achieve in Maths

One thing that I notice about mathematics in schools is that many pupils are afraid of "getting it wrong"; getting the wrong answer or saying the wrong thing. Getting things wrong is actually an extremely important part of maths. If you get something wrong in maths but then understand what went wrong, why it went wrong, and what you should have done instead, then this is just as good, if not better, than getting it right in the first place. The only exception to this rule is when you are sat in your examination, because if you get it wrong in your exam it is too late to learn what you should have done instead. The aim is to make every mistake possible before you get to your exam, so that by the time you are sitting your exam, you know all the different things that can go wrong, and you know how to avoid them. Once you realise that maths is not about getting things right all the time; but about trying different ways to do things and seeing what works, then a lot of pressure is released, you can start to enjoy your learning without the fear of not doing something right!

All you have to do is do something, do something that feels right, or feels like it might be right. Go with that and then analyse it after, to see how far from the correct answer you are and what you should have done differently. You must keep trying to improve your approach each time, or changing your method completely if needs be. If you get an answer right straight away then that is great, but you will not learn anything from that question, and it will not help you to progress. The only time that we really learn is when we make our own mistakes.

One detrimental idea to becoming "good" at maths is to think that it is acceptable to not be good at maths! In our society today there are people that are not good at maths, and they do not want to be good at it. I have heard many pupils, and even many teachers, stating that they "cannot do Maths". It is upsetting to me that a teacher would proudly announce to their class that they were "no good at Maths, and it never did them any harm". Teachers are supposed to be role models, supposed to try harder when things are difficult and not just admit defeat. An observation that I have made is that, in general; adults would be ashamed or embarrassed to admit that they could not read (as one example), but some adults seem almost proud to proclaim that they do not get maths. I think that because of this stigma, children are missing out on opportunities to shine at mathematics. Even the media portrays the subject as out of touch with everything else. The truth is, maths is completely in touch with almost everything else. It underpins everything that we see around us. Without knowledge and understanding of mathematics, our economy would simply not function and we would not have the incredible technology that we can enjoy today. English is a language that everyone wants to be good at, maths is a universal language that everyone should want to be good at. One can argue about semantics in language, and struggle over what is "correct". In the language of mathematics, there can be little room for arguments, we are talking about fundamental truths.

If you take one object and add another object; then you have two objects, regardless of what your words for "one", "two" or even "object" are. If you have seven rows of five objects; then whichever way you look at it, you have thirty-five objects. The building blocks of mathematics are real, each new mathematical fact or idea that you learn is just as real as the one before it (because mathematicians prove them rigorously). There are no nasty surprises or uncertainties, just some extra things to try and get your head around; so that you can try to understand the rationale and the reason, because it will be there somewhere.

My advice is to see mathematics as a quest for discovery; question what you are told, ponder how it works, and try to get your brain to understand why. Once you understand why something works then it is time to practice the skills. You may understand and be able to carry out the skills one day, but without purposeful practice of the techniques; unfortunately, your brain will forget and become "rusty". To keep sharp, you need to keep reminding yourself of what you have learnt. This is why you can only get very good at maths by actually doing maths. Only you have the power to choose if to really "do" maths. To really engage your brain and participate in the problem-solving activity presented. Only you have the capability of improving your mathematical ability, no one else can simply transfer the knowledge and skills into your head. When it comes to learning mathematics, you have a huge responsibility. You have to say "yes" to getting involved, and give it everything that you have got.

IV. Action Steps to Achieve in Maths

We have talked a great deal about what it means to be good at maths, and the theory behind it. In this chapter I want to give you some tangible action steps that you can begin to take straight away, to begin your journey of achieving in maths.

The first step is to listen intently in maths lessons or to your maths tutorials (online or otherwise). Try your best to follow and make sure that you ask questions when things are not clear to you. Either post a comment on the YouTube video that you are watching or ask your teacher (if you don't want to ask in front of other people then wait behind after class to ask). The key concept here is not to leave any gaps in your knowledge, it does not actually matter how you go about this. Remember, the emphasis is not on getting the answer, it is on understanding the material. Google can be an extremely useful tool; try searching for the topic that you are struggling with and read different sources until one of them makes sense to you.

Once you feel that you understand the technique that you are learning, it is time to practice it as much as possible. This is the reason why you are given homework at school / college, so that you can practice what you have learnt in your own time, and at your own pace. It is very important to take your homework seriously. If you do not think that you have been given enough homework to practice the technique thoroughly, then use Google or a textbook to find extra questions for you to work on.

One of the challenges of examinations in mathematics is identifying what the question is asking you to do. If you only ever practice one topic at a time, then you will struggle with this. Eventually, you will need to practice questions from a range of topics at the same time. Revision in maths is about preparing yourself for the exam, by putting yourself in a similar position to that which you will be in when you sit the exam. Ideally, you need new questions that you have not seen before, with no warning of the topics. This is why I have created a set of completely new and original exam style questions, these can be found in the next chapter. I believe that the best revision practice that you can do is to get a past exam paper of the subject that you will be sitting, and try to attempt it yourself, with no help. This can be very daunting at first, so you can use your notes or other sources to help you when you are stuck at the beginning.

It is imperative that by the time the exam comes, you can sit down and complete a past exam paper under exam conditions, in the allocated time. Therefore, you need to build up to this yourself. Find somewhere that you will not be disturbed. Set your timer for the allocated time, and attempt a past exam paper, pretending that it is the real thing. After you have done this, you can gain a great deal of value by comparing what you did to model solutions. This is where I recommend that you use videos on YouTube that talk you through past exam papers (I have a growing collection of these videos on my channel – Achieve Maths). Give yourself marks for the ones that you got right, and follow the explanations for those that you did wrong. Try to see what misunderstanding caused you to go wrong and learn what you should have done instead. You can make a note of this as a topic that you need to do more practice on. Focus on your weakest areas because if you are already good at a certain type of question, there is no point in wasting valuable time on doing more of this topic. Instead, you can improve your understanding of a topic that you find difficult, doing this will actually have an impact on the number of marks that you can collect from this type of question, next time you do an exam paper.

An alternative approach to using past paper walk through videos as complete solutions, is to pause the video on each question. Attempt it yourself then play the video to see how it should be done. This method is useful, but the disadvantage is that you are not timing how long it takes you to complete the paper. Even if you can do every possible question, you could still fall short of your desired grade if you run out of time and fail to complete the whole paper.

You need to be comfortable with what kind of a performance in the paper will get you the grade that you are aiming for. If, for example, you are aiming for a grade B in GCSE Edexcel Mathematics; you should look up past grade boundaries to find that you will need to score roughly 48%. There is always a chance, that even though you prepare as well as you can for your exam, things do not go as well as you would have hoped on the day. In your exam preparation, you must aim higher than the boundary level for your grade. You cannot just aim to scrape through the grade boundary and hope for the best. My advice would be to aim for at least 10% higher than the percentage score that you need. This way you will leave yourself some room for errors and mistakes. If in your preparation for the exam you are managing to score say 60%, even though you may only need 50%. You can be more relaxed in your actual exam, and know that even if things go a little worse than they have done in your practice, you should still achieve the grade that you want.

Start your preparation early; there is no such thing as doing too much revision. Just make sure that the revision and practice that you are doing is efficient, by using the techniques discussed in this chapter. Imagine yourself succeeding in achieving the grade that you are striving for in Mathematics. Once you achieve this grade, it is yours forever. No one can take that success away from you, you will have worked hard, and you will have earned it. I wish you every success in your venture and I believe in you. The fact that you have read this book has shown your determination. Take this determination with you to your studies, and good luck in the challenging yet highly rewarding journey of achieving in maths.

The following chapters are based around mathematical questions and how to approach them. This will give you the chance to put all the ideas discussed in this book into practice.

V. Maths Questions that will make you think

This section contains functional exam style questions that I have created. They would probably be graded as grade A or A* in difficulty in a Mathematics GCSE exam. I have intentionally designed these questions to be challenging so do not be discouraged if you find them difficult. These questions require an understanding of the underlying mathematics involved. You will need dedication to keep trying at a problem and to work through it systematically. Not only do you need to have the subject knowledge of the topic being tested, you also need to be able to apply it to new situations that you have not seen before. My aim is that by preparing you in this way, you will think creatively about tackling a maths problem. Some of the questions will help you to understand more about how the mathematical skills that you have developed can be applied to real life situations.

I have provided hints for some of the questions and the solutions are given in the next chapter. Make sure that you only look at the answers when you are convinced that you have reached the correct answer, or if you are completely stuck and there is nothing else that you could try yourself.

1.

Question:

A man is stood 25m away from the centre of a building. He plans to use a protractor, ruler and trigonometry to estimate the height of the building. He holds the protractor so that the bottom is parallel to the floor and he points the ruler from his eye level to the top of the building as shown below.

The man is six foot and three inches tall, he has measured that the distance from the top of his head to his eye level is 14.5cm. The reading from his protractor and ruler set up is shown below.

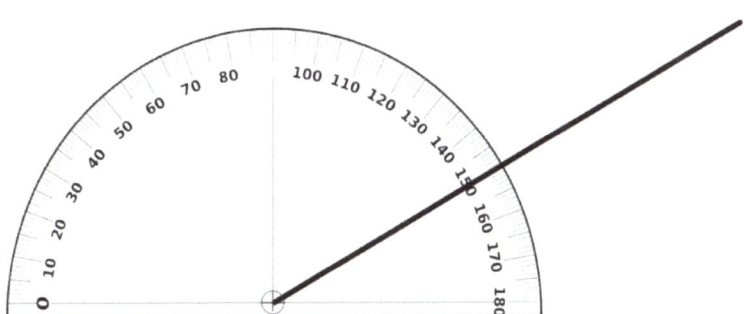

The man has used the internet to find that 1 inch is equal to 2.54cm and he knows that there are twelve inches in one foot. Estimate the height of the building.

2.

Question:

If you have three sticks of lengths 5cm, 8cm and 10cm, can you make a right-angle triangle? Explain your answer with mathematical justification.

Hint:

Use the result of Pythagoras' Theorem.

3.

Question:

Abbie starts with £35 then gets £1.50 a day, Bill starts with £11 then gets £2 a day.

    a) After how many days will they have the same amount of money?

    b) After how many days will Bill have 20% more than the amount that Abbie has?

c) Catherine starts with the average of Abbie and Bill's starting amounts. She flips a fair coin each day, if it is heads then she gets £1.50 and if it is tails she gets £2. About how much would you expect Catherine to have after 20 days?

d) Explain in words if your answer for part c) is reliable and give reasons for this.

e) If Abbie has amount A on one day and Bill has an amount B on the same day, derive a formula for the amount C, where C represents the amount that you would expect Catherine to have on that day.

4.

Question:

a) How many times bigger does the area of a rectangle get if all of its sides get twice as big? Give a reason for your answer in terms of algebra.

b) How many times bigger does the area of a rectangle get if each side is multiplied by n?

Hint:

For part a), draw out a rectangle and find its area, then draw out a new rectangle with sides that are double the original and find the new area.

5.

Question:

Mason is trying to make a mixture which is made up of 2.5% acid and the rest water. He has got diluted acid to use which is half water and half acid. How much water and how much of the diluted acid should be combined to make a 300ml mixture?

6.

Question:

Two dogs have been left outside a supermarket while the owners go inside, one dog is attached to a post and the other is attached to a railing. They are attached by their leads which are extendable. The dog attached to the post is on a lead which extends to 5m and the dog on the railing is attached by a lead which extends to 3m.   The railing is 6m long and is only attached to the floor at both ends, this means that the lead can slide along it from end to end as shown on the diagram below.

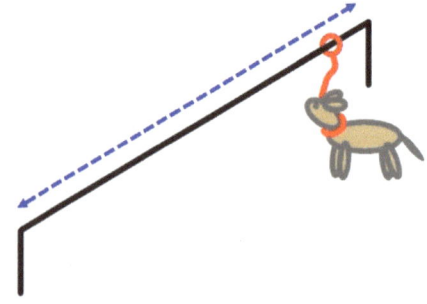

Show, on the diagram below, all of the possible places that each dog can go. Use an appropriate key and explain whether or not the dogs can get to each other.

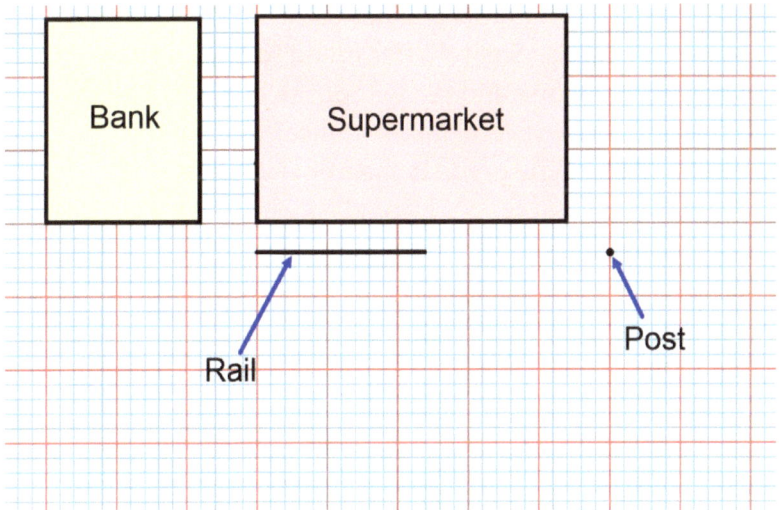

7.

Question:

Is it quicker to travel at an average speed of 60mph for one hour, or an average speed of 55mph for 40 minutes and an average speed of 68mph for 20 minutes?

8.

Question:

Kelly is very good at maths and has come up with a trick to impress her friends. She tells her friend to think of a number between one and ten without telling her. She then tells them to apply several mathematical steps to the number and correctly predicts their final answer.

The steps that she tells them to do to their number are as follows:

First add on three to your number, then square the result, then multiply this by two, then subtract eighteen. Next divide by the number that you first started with and then add on four. Now divide by two and finally, take away the number that you first started with.

Kelly always predicts that the final answer is 8.

a) Work through the steps using algebra and explain whether or not this will always work.

b) Why does Kelly tell her friends to start with a number between 1 and 10?

9.

Question:

You are trying to send an important parcel to a friend, you want to know the probability that the parcel will arrive to him safely. You have read online that 0.5% of parcels get "lost in the post". Your friend informs you that if he is not in at the house when a parcel is delivered, then it is left on the porch. He also tells you that out of the 10 parcels that he has had delivered this year, 3 of them had to be left on the porch. You are worried that if the parcel is left on the porch it might get stolen and you think that there is a 5% chance of this happening.

    a) Estimate the probability that the parcel makes it to your friend safely.

    b) What makes this estimate unreliable?

10.

Question:

Prove that angles in a triangle always add up to 180°.

Hint:

There are 360° degrees around a point (in a full rotation) and exterior angles of a convex polygon add up to 360°.

11.

Question:

Ahmed and Britney go into a shop. Ahmed buys three apples and two Mars bars, Britney buys one apple and five Mars bars. They are in a rush so they do not check the prices of the items. Ahmed pays £1.85 and Britney pays £2.61.

    a) Carl Sees Ahmed and Britney as they leave the shop and he tells them that he is going to the shop to buy three mars bars. How much would this cost him?

    b) Carl changes his mind when he is in the shop and he buys four apples and three Snickers bars instead. This costs him £2.71. How much was one Snickers bar in the shop?

12.

Question:

Doug invested an amount of money in some shares. He had to pay a charge of £30 to buy the shares, but there were no other fees. After three years, he has £1,749.33 worth of shares. In the first year of his investment the shares rose by 8%, in the second year they fell by 5% and in the third year they rose by 10%.

a) How much money did Doug have to invest in the shares before the charge was taken?

b) Doug sells two thirds of the shares, there is a £25 charge for selling the shares. Doug purchases a television for £350 and then invests the remaining money in a savings account. The savings account guarantees to pay 2% interest per year (compound interest).

   Doug predicts that, on average, the shares will gain in value by 4% each year. Predict how much his investments (shares and savings) will be worth after twelve years, if he is correct.

c) How much extra would Doug save if he invested the amount needed for the television into the savings account instead?

d) In twelve years' time, Doug's shares are actually worth £769.74. Use trial and improvement to find the actual average percentage that the shares grew by each year (to one decimal place).

e) Doug tells his friend about the percentage return that he has achieved in part d). He tells his friend that because the average percentage growth rate that he earned on the shares is higher than the interest rate in the savings account, he is always better off investing in shares. Is this correct? Give a reason for your answer.

VI. Model Solutions to Questions

1.

Worked Solution:

This question gives us measurements in a range of different units, we need to convert all lengths to a common unit of measurement to make calculations easier. Metres seems most sensible for the height of a building so we will work in metres throughout the question.

We are told that the man is 6 foot and three inches tall, since there are 12 inches in 1 foot the height of the man is 6 x 12 + 3 = 75 inches. There are 2.54 centimetres 1 inch so the man's height is 75 x 2.54 = 190.5cm. The height from the ground to the man's eye level will be his height minus the distance from the top of his head to his eye level, i.e. 190.5cm – 14.5cm = 176cm = 1.76m.

Trigonometry works with right-angle triangles so we will need to use the information that we have been given to establish a right-angle triangle. We can treat one corner of the triangle as the angle measured with the protractor, another corner as the top of the building and the final corner as the right-angle produced as shown in the diagram below.

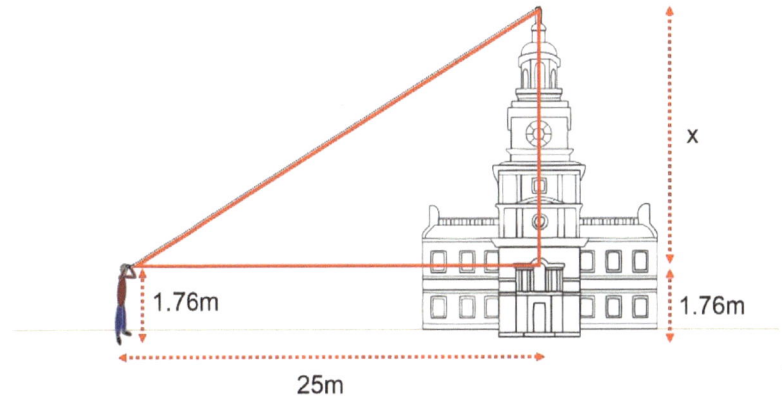

1.76m

x

1.76m

25m

The protractor reads 150°, however this is the obtuse angle, we want the acute angle which is 30° as shown on the diagram below (180° - 150° = 30°). So 30° is the angle between the line parallel to the floor (at 1.76m from the ground) and a line to the top of the building. The task now is to find the missing length x, which will be added to 1.76m to find the total height of the building.

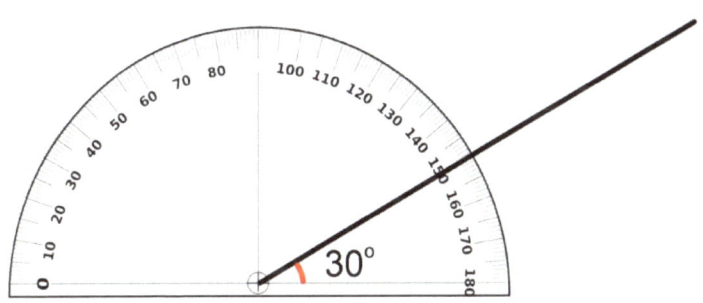

30°

The resulting triangle is shown in the diagram below. We do not need to know the hypotenuse of the triangle (the diagonal distance from the man's eyes to the top of the building), we need to know the height of the triangle (the length opposite the 30° angle). We know the length adjacent to the 30° angle is 25m. This means that we can use the trigonometric equation for tangent to find x.

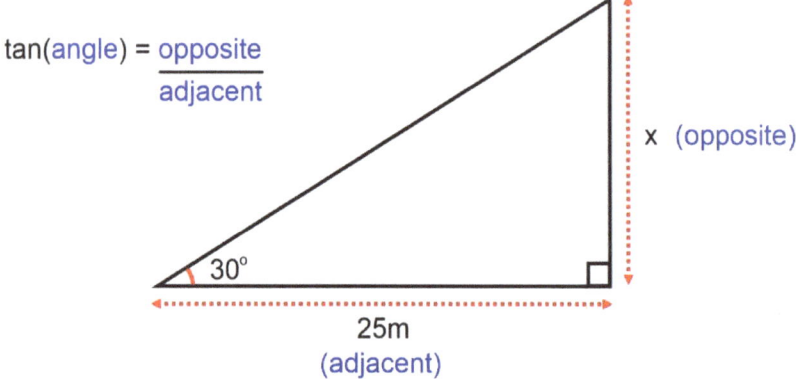

Substituting our known values and unknown value (x) into the formula gives us the following:

$$\tan(30°) = x / 25$$
=>   25 x tan(30°) = x
=>   14.43375673 = x
=>   x = 14.43m (to 2 decimal places)

The height of the building is 1.76m + 14.43m = 16.19m so our estimate for the height of the building is 16m to the nearest metre.

2.

Worked Solution:

Pythagoras' Theorem states that in a right-angle triangle, the sum of the squares of the two shorter lengths gives you the sum of the square of the hypotenuse. In algebra, this can be written as $a^2 + b^2 = c^2$ where c is the hypotenuse and a and b are the other two sides (either way around). This means that this is only true if our triangle is a right-angle triangle, i.e. if $a^2 + b^2$ does not equal c^2 then our triangle is not a right-angle triangle.

The hypotenuse is always the longest length so we can let 10cm = c and let 5cm and 8cm be a and b (either way around).

$a^2 + b^2 = 5^2 + 8^2 = 25 + 64 = 89$, whereas $c^2 = 10^2 = 100$
This means that $a^2 + b^2$ does not equal c^2 and hence we cannot make a right-angle triangle with our three sticks.

3.

Worked Solution:

a) Let x = the number of days that have passed when Abbie and Bill have the same amount of money.

Abbie has amount (£):	35 + 1.5x
Bill has amount (£):	11 + 2x

We know that these two amounts are the same so we can make the two expressions equal each other and we get an equation to solve as follows:

$$35 + 1.5x = 11 + 2x$$

$\Rightarrow \qquad 24 + 1.5x = 2x$

$\Rightarrow \qquad 24 = 0.5x$

$\Rightarrow \qquad 48 = x \quad \text{i.e. } x = 48$

Hence after 48 days Abbie and Bill will have the same amount of money.

Check:

After 48 days:
Abbie will have $35 + 1.5 \times 48 = 35 + 72 = £107$
Bill will have $11 + 2 \times 48 = 11 + 96 = £107$, so we know that our answer of 48 days is correct.

b) Let y = the number of days that have passed when Bill has 20% more than the amount that Abbie has, this means that Bill has 120% of Abbie's amount (i.e. 1.2 times it).

Abbie's amount will be $35 + 1.5y$ and so Bill will have $1.2 \times (35 + 1.5y)$.

We also know that Bill will have amount $11 + 2y$, this means that we can equate these two expressions and we get the following equation to solve:

$$1.2 \times (35 + 1.5y) = 11 + 2y$$

Expanding out the brackets on the left-hand side of the equation gives:

$$42 + 1.8y = 11 + 2y$$
$$\Rightarrow \quad 31 + 1.8y = 2y$$
$$\Rightarrow \quad 31 = 0.2y$$
$$\Rightarrow \quad 155 = y \qquad \text{so } y = 155$$

Therefore, after 155 days Bill will have 20% more than Abbie.

Check:

After 155 days:
Abbie will have 35 + 1.5 x 155 = 35 + 232.5 = £267.50
Bill will have 11 + 2 x 155 = 11 + 310 = £321
£267.50 x 1.2 = £321, so Bill has 20% more than Abbie (120% of Abbie's amount)
Hence our answer of 155 days is correct.

c) Abbie starts with £35 and Bill starts with £11, the average of these starting amounts is £35 plus £11 divided by 2, i.e. £46 / 2 = £23. Each day Abbie has a 50% chance of getting £1.50 and a 50% chance of getting £2. This means that on average, for half the days she should get £1.50, and for the other half of the days she should get £2. Therefore, the most likely option is that she gets ten days of £1.50 and ten days of £2, this gives a total amount of:

£23 + 10 x £1.50 + 10 x £2 = £58

Hence, after 20 days we expect Catherine to have £58 (but the actual amount that she has could well be different to this).

d) The more times that you flip a coin, the closer to half heads and half tails you will get. Twenty is too small a number to be confident that your number of heads and tails from your coin flips will be roughly equal. The fact that we are only thinking about twenty days (i.e. twenty coin flips) means that our expected value from part c) is unreliable. Since there is an element of luck involved, it is impossible for our expected value to ever be completely reliable, in terms of the result that we will actually get.

e) Since we cannot be sure whether Catherine will get £1.50 or £2 on a day, we can only give an average (or expected value) of what she gets on a day. The average amount that she will get per day is £1.50 plus £2 divided by 2, i.e. £3.50 / 2 = £1.75. Catherine starts off with £23, so her expected amount after n days (C) will be:

C = £23 + £1.75 x n

The amounts that Abbie and Bill will have after n days, in formula form, is as follows:

A = £35 + £1.50 x n

B = £23 + £2 x n

By comparing the formulae, we can see that C is the average of A and B, i.e. that (A + B) / 2 gives C. This means that a formula for the amount C, where C represents the amount that Catherine would expect to have on a day where Abbie has amount A and Bill has amount B is:

C = (A + B) / 2

4.

Worked Solution:

a) You could start by drawing out a rectangle with sides of any length, for example a 3 by 4 rectangle. If you were to double all the sides, then you would have a 6 by 8 rectangle. The original rectangle has an area of 3 x 4 = 12 square units and the new rectangle has an area of 6 x 8 = 48 square units. In this case the area has become 4 times bigger since 12 x 4 = 48.

To give an answer in terms of algebra, we will need to start with a rectangle with sides of any length possible, i.e. an x by y rectangle, where x and y can be any numbers. If we double every side, then our new rectangle will be a 2x by 2y rectangle. The area of the original rectangle will be xy square units and the area of the new one will be 2x x 2y = 4xy square units. 4xy is four times xy, this means that regardless of the values of x and y, the rectangle area will get four times bigger if we double every side.

b) If we start with a general rectangle of size x by y then multiply all sides by n, we will get a rectangle of size nx by ny. The new rectangle will have area nx x ny = n^2xy which means that the area of a rectangle will get n^2 times bigger if every side is multiplied by n.

5.

Worked Solution:

Mason wants 2.5% of a 300ml mixture to be acid, 2.5% of 300ml = 7.5ml (2.5% is 0.025 as a decimal and 0.025 x 300 = 7.5). To get 7.5ml of acid, he will need to use double this amount of the diluted acid, since only half of the diluted acid is actually acid. This means that he will need 15ml of the diluted acid. The remainder of the mixture can be made up of water. Hence, he will need 300ml – 15ml = 285ml of water.

6.

Worked Solution:

We can figure out the scale used on the diagram by using the rail, which is 6m, as shown below:

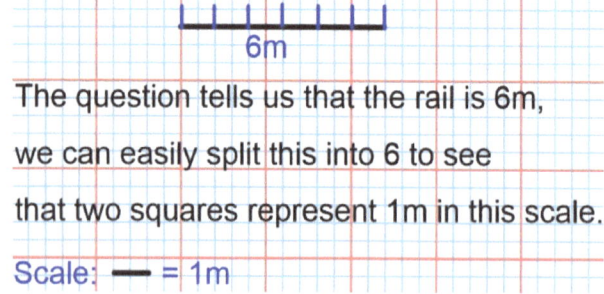

6m

The question tells us that the rail is 6m, we can easily split this into 6 to see that two squares represent 1m in this scale.

Scale: ▬ = 1m

The dog attached to the post can go to any position up to 5m (10 squares) away. We can get all of the points that are exactly 10 squares away from the post by drawing a circle (using a compass) with the post as the centre, and with the radius as 10 squares. Any position inside this circle is less than 5m from the post.

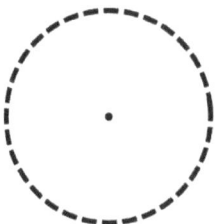

The dog attached to the rail can go to any position up to 3m (6 squares) from the rail. Each end of the rail can be thought of as a post and so, as above, we can draw circles of radius 6 squares, with the centres as each end of the rail. The furthest that the dog can get from the middle of the rail is 3m away, on a line perpendicular to it (at 90° to it). This means that two lines, parallel to the rail, one 6 squares above it and one 6 squares below it, give all the points that are as far as possible away from the rail (above and below it). The resulting area is within all of these construction lines. The area gives all of the positions that the dog can get to, from each side of the rail, and from the middle section of the rail. The resulting area is shown below (with and without working out lines).

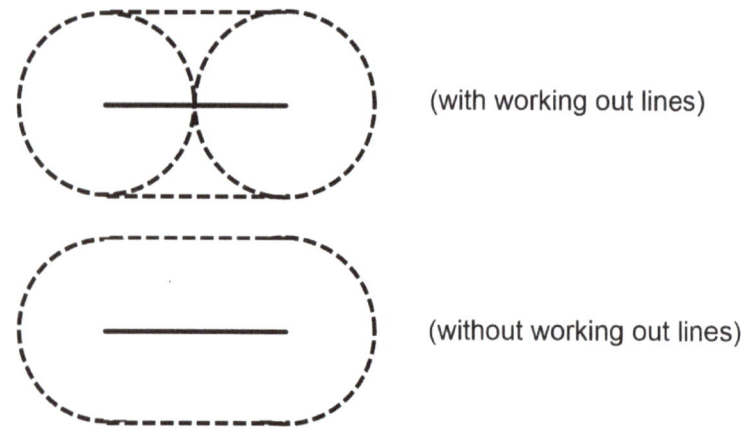

(with working out lines)

(without working out lines)

The diagram shows the supermarket and a bank, the dogs will not be able to walk through walls so our areas must stop immediately at the boundaries of each building. If we shade each of the areas that the two dogs can get to then we see that the two areas overlap. This means that the dogs can get to each other. This can be made clear with shading and by labelling the diagram as shown in the solution below.

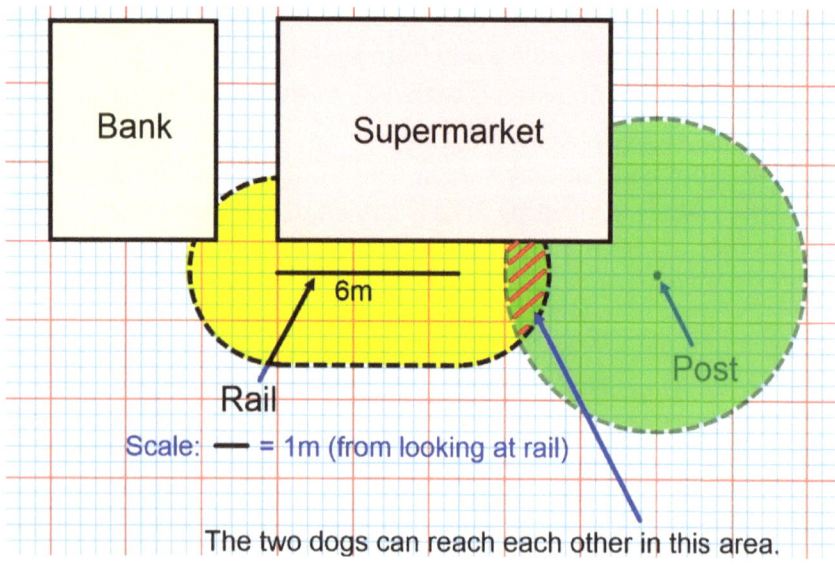

Bank

Supermarket

6m

Rail

Post

Scale: ▬ = 1m (from looking at rail)

The two dogs can reach each other in this area.

7.

Worked Solution:

We can split an hour into 60 minutes.

For the simple case of traveling at 60mph we have the same average speed for all of the 60 minutes.

For the second case, we have an average speed of 55mph for 40 of the 60 minutes, and an average speed of 68mph for 20 of the 60 minutes. We can find the weighted average of these average speeds as follows:

$$40/60 \times 55\text{mph} + 20/60 \times 68\text{mph}$$
$$= 2/3 \times 55\text{mph} + 1/3 \times 68\text{mph}$$
$$= 59.3\text{mph which is less than 60 mph}$$

Therefore, it is quicker to travel at an average speed of 60mph for one hour.

8.

Worked Solution:

a) Let's call the number that one of Kelly's friends has thought of "x", this is because we do not know what it is (it could have been any number between 1 and 10 so it is an unknown).

Step 1: Add 3
Result after Step 1: x + 3

Step 2: Square it
Result after Step 2: $(x+3)^2 = x^2 + 6x + 9$

Step 3: Multiply by 2
Result after Step 3: $2(x^2 + 6x + 9) = 2x^2 + 12x + 18$

Step 4: Subtract 18
Result after Step 4: $2x^2 + 12x = x(2x + 12)$

Step 5: Divide by the number that you first started with (divide by x)
Result after Step 5: $2x + 12$

Step 6: Add 4
Result after Step 6: $2x + 16 = 2(x + 8)$

Step 7: Divide by 2
Result after Step 7: $x + 8$

Step 8: Subtract the number that you first started with (subtract x)
Result after step 8: 8

We can see that x has disappeared from our algebra. It does not matter what the value of x was (what the number that we started with was), we will always be left with just 8 on its own after all the steps have been completed. Kelly's trick will always work, as long as her friends carry out the mathematical steps correctly!

b) Kelly tells her friends to start with a number between 1 and 10 so that the calculations are easier for them, this means that it is less likely for them to make a mistake. Hence, they are more likely to be left with the correct answer of 8 at the end of the steps.

9.

Worked Solution:

a) Firstly, we must understand all the possible ways that the parcel can arrive safely to your friend. The easiest way to see this is to draw a tree diagram of all of the possibilities as shown below:

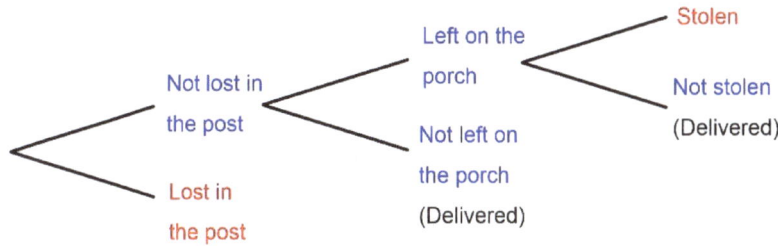

We can see that there are two options for safe delivery. The first option is for the parcel not to be lost in the post and then not left on the porch (i.e. delivered). The second option is for the parcel not to be lost in the post, for it to be left on the porch, but for it not to be stolen.

We now consider the probability of each event and we want to use the same format for each probability (i.e. percent, decimal or fraction), decimals will be most useful in this case. 0.5% of parcels get lost in the post, 0.5% as a decimal is 0.5/100 i.e. 0.005. Since it can either get lost in the post or not lost in the post (and there are no other possibilities) these two probabilities must add up to 1 (i.e. one or the other happening is certain). So, the probability of the parcel not getting lost in the post is one minus the probability of it getting lost in the post, i.e. 1 – 0.005 = 0.995. If 3 out of 10 parcels are left on the porch, then we can estimate from this that the probability of a parcel being left on the porch is 3/10 = 0.3. This means that the probability of the parcel not being left on the porch is 1 – 0.3 = 0.7 (since we are assuming that these are the only two possibilities). We are going to base our calculation on our idea that there is a 5% chance of the parcel being stolen from the porch, 5% as a decimal is 0.05.

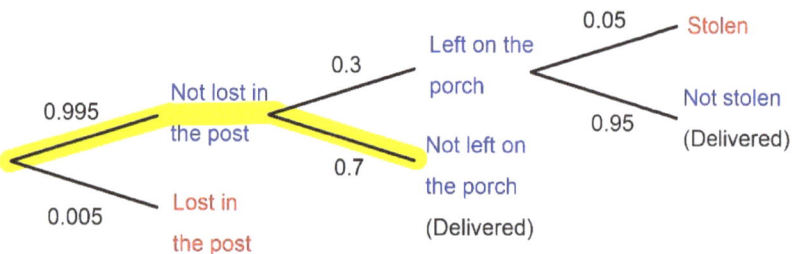

The diagram above shows the tree diagram with the probabilities labelled on and the first possible way that the parcel can be delivered is highlighted (option 1).

The diagram below highlights the second possible way that the parcel can be delivered (option 2).

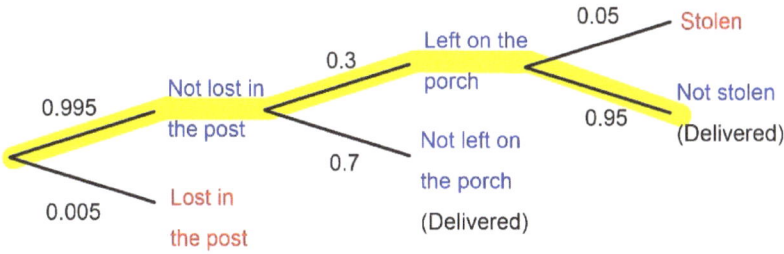

If we want to know the probability of one or more independent, mutually exclusive events (events that cannot happen at the same time and do not affect each other) happening, then we can multiply the probabilities.

For option 1, we want the parcel not to be lost in the post and for it not to be left on the porch. We can therefore multiply the probability of it not being lost in the post by the probability of it not being left on the porch to get the probability of it being safely delivered in option 1. Hence the probability of it being safely delivered in option 1 is 0.995 x 0.7 = 0.6965.

For option 2, we want the parcel not to be lost in the post, for it to be left on the porch and for it not to get stolen from the porch. We can therefore multiply the probability of it not being lost in the post by the probability of it being left on the porch by the probability of it not getting stolen from the porch, to get the probability of it being safely delivered in option 2. Hence the probability of it being safely delivered in option 2 is 0.995 x 0.3 x 0.95 = 0.283575.

If we want to know the probability of one or more independent, mutually exclusive events happening, then we can add the probabilities. This means that to get the probability of the parcel being safely delivered, we can add the probability of it being delivered as in option 1 to the probability of it being delivered in option 2. Thus, our estimate of the probability that the parcel is safely delivered is 0.6965 + 0.283575 = 0.980075 i.e. 0.98 (98%) to 2 decimal places.

b) It does not consider the possibility of other events happening, for example, the parcel being delivered to the wrong house by mistake.

All of our probabilities are estimates because we are using estimates for the probability of each event, e.g. we estimated that because 3 out of 10 parcels were left on the porch, this means that there is a 30% chance of the next parcel being left on the porch. The sample size of 10 is particularly small so the estimate of 30% is unreliable.
The probability of the parcel being stolen from the porch could be anything, we are just basing our estimate on our idea of 5%, which might be wrong.

10.

Worked Solution:

To prove that the angles in any triangle add up to 180°, we must consider a triangle with interior angles A, B and C. Where A, B and C can be any size, as long as they still make a triangle. If we can prove the result by using the general letters, then we know that it is true for whatever numerical value that A, B and C may take.

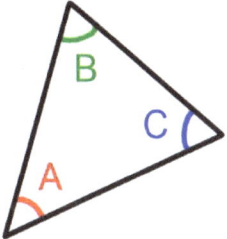

We can start by drawing out an arbitrary triangle and labelling the angles as shown above. Now we can adapt our diagram to investigate the angles further. If we extend each side of the triangle as shown below, then we produce the exterior angles of the triangle, we can call the exterior angles of the triangle A', B' and C'.

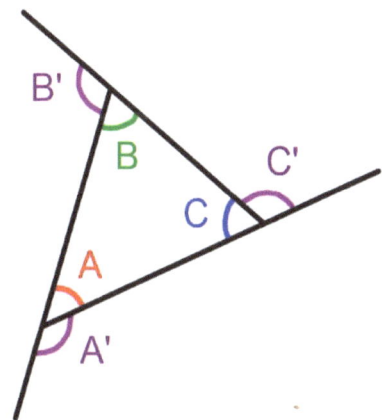

We know that the exterior angles of a convex polygon add up to 360°. This means that the exterior angles of a triangle add up to 360° (you can visualise this by imagining that you walk around a triangular shaped building, when you get back to your starting point, your body has turned around by exactly 360°). With reference to our diagram, this means that $A' + B' + C' = 360°$.

We can think of each corner of the triangle as a single point and we know that there are 360° around a point (in a full turn). This means that in half a turn (i.e. on a straight line), there are 180°. From our diagram, we can see that as a result of this we have the following equations:

$$A + A' = 180°, B + B' = 180° \text{ and } C + C' = 180°.$$

If we add these three equations together we get the following result:

$$A + A' + B + B' + C + C' = 180° + 180° + 180°$$
$$\Rightarrow \quad A + B + C + A' + B' + C' = 540°$$
$$\Rightarrow \quad A + B + C = 180° \quad \text{(since } A' + B' + C' = 360°\text{)}$$

Since A, B and C are the angles of any given triangle, this has proved our result, that the angles in a triangle add up to 180°.

11.

Worked Solution:

a) Let M be the price of a Mars bar and A be the price of an Apple. We can then write expressions for the total cost of the items bought, and equate them to the amounts paid by Ahmed and Britney as follows:

Ahmed: 3 x A + 2 x M = £1.85

Britney: 1 x A + 5 x M = £2.61

This can be written as simultaneous equations as follows (ignoring the units for now):

$$3A + 2M = 1.85 \quad \text{(equation 1)}$$
$$A + 5M = 2.61 \quad \text{(equation 2)}$$

If we multiply the whole of equation 2 by 3 then we will have common terms of 3A in both equation 1 and equation 2:

(equation 2) x 3   3 x A + 3 x 5M = 3 x 2.61

=>   3A + 15M = 7.83   (equation 3)

Now, if we subtract equation 1 from equation 2, we will cancel out the 3A term (this is called the elimination method for solving simultaneous equations).

$$3A + 15M = 7.83 \quad \text{(equation 3)}$$
$$- \ 3A + \ \ 2M = 1.85 \quad \text{(equation 1)}$$

$$= \quad \ \ \ 13M = 5.98$$

We can now solve the resulting equation, because it only has one unknown value. In words, the equation is saying that thirteen Mars bars cost £5.98. We can therefore divide £5.98 by 13 to find the cost of one Mars bar. This gives that one Mars bar costs £0.46, and in algebra, M = 0.46. To find the value of A, we must substitute the value of M into one of our original equations (equation 1 or equation 2) and solve the resulting equation. Using equation 2:

$$A + 5 \times 0.46 = 2.61$$
$$=> \qquad A + 2.3 = 2.61$$
$$=> \qquad A = 0.31$$

We can check this answer by substituting our values of A and B into our other original equation (equation 1), to make sure that the left-hand side of the equation equals 1.85.

Check:

$$3A + 2M = 3 \times 0.31 + 2 \times 0.46 = 0.93 + 0.92 = 1.85$$

Our check confirms that our values for A and B are correct and hence an apple costs £0.31 and a Mars bar costs £0.46

Carl plans to buy three mars bars, so this will cost him 3 x £0.46 = £1.38.

b) We know from part a) that the cost of an apple is £0.31. This means that four apples will cost 4 x £0.31 = £1.24, we can subtract this from the total cost (£2.71) to find that three Snickers bars cost £2.71 - £1.24 = £1.47. We can then divide the cost of three Snickers bars to find the cost of one of them, i.e. the cost of one Snickers bar is £1.47 / 3 = £0.49.

12.

Worked Solution:

a) Let M be the amount of money that Doug had to invest. He was charged £30 to buy the shares, so the amount that was actually invested in the shares was M − 30.

This amount rose by 8%, which is equivalent to multiplying the amount by 1.08 (since 100% + 8% is 1.08 as a decimal). It then fell by 5%, which is equivalent to multiplying the amount by 0.95 (since 100% - 5% is 0.95 as a decimal). Then it rose by 10%, which is equivalent to multiplying the amount by 1.1 (since 100% + 10% is 1.10 as a decimal). This means that the amount of money that Doug has at the end of the third year is:

$$(M - 30) \times 1.08 \times 0.95 \times 1.1 = (M - 30) \times 1.1286$$

We know the amount of money that Doug has after three years is £1,749.33, we can equate this to our above expression and solve the resulting equation to find M as follows:

$$(M - 30) \times 1.1286 = 1,749.33$$

=> $(M - 30) = 1,550$

=> $M = 1,580$

Hence the amount of money that Doug had to invest in the shares before the charge was taken was £1,580.

b) Doug has £1,749.33 worth of shares, one third of this amount is £1,749.33 / 3 = £583.11, two thirds is therefore 2 x £583.11 = £1,166.22. When Doug sells two thirds of the shares (£1,166.22 worth), he is charged £25, so he is left with £1,166.22 - £25 = £1,141.22, to invest in the savings account.

The amount in the savings account will get 1.02 times bigger each year since this is equivalent to adding 2% to your current 100% each year (100% + 2% = 102%, which is 1.02 as a decimal and once in the form of a decimal it can be used as a multiplier). Since there are twelve years, we will multiply our original amount by 1.02 twelve times, which is simplified by using the power of twelve. This gives us the equation:

$$£1,141.22 \times 1.02^{12} = £1,447.342901$$
$$= £1,447.34 \text{ (to the nearest penny)}$$

Doug predicts that, on average, the shares will gain in value by 4% each year. We can use the same technique as above to increase the original shares value (£583.11) by 4% each year, i.e. by multiplying by 1.04^{12} as follows:

$$£583.11 \times 1.04^{12} = £933.577897$$
$$= £933.58 \text{ (to the nearest penny)}$$

This means that his investments (shares and savings) will be worth £933.58 + £1,447.34 = £2,380.92 to the nearest penny, if Doug's prediction is correct.

c) The television cost £350, if Doug were to invest this money in the savings account for twelve years instead then he would have £350 x 1.02^{12} = £443.88 (to the nearest penny). This means that the extra amount that he would have saved is £443.88 - £350 = £93.88.

d) In order to use trial and improvement, we must set up an equation that needs to be solved. We know that at the beginning of the twelve years', the shares are worth £583.11, and at the end of the twelve years they are worth £769.74. Let P represent the actual average percentage that the shares grew by each year. In each year, the shares will get (1 + P) times bigger (for example, if the growth rate was 5% then the shares would become 105% of their original worth, which is the same as getting 1 + 0.05 = 1.05 times bigger). Since there are twelve years, we will multiply our original amount by (1 + P), twelve times, which is simplified by using the power of twelve. This gives us the equation:

$$583.11 \times (1 + P)^{12} = 769.74$$

The easiest way to do trial and improvement is with a table. We need one column for the value of P that we are trialling (our guess), one column which shows us the answer to the left-hand side of the equation (that we get if we use that particular value of P). And we need one column to say if our answer is higher or lower than the right-hand side of the equation (what we are aiming for). If our answer is too high, then we will need to make our next guess for P smaller and if our answer is too low, then we will need to make our next guess for P bigger. For our first guess, we can start with 4% (0.04), since this was

Doug's prediction (but it does not matter what value you start with, it will just take you longer if your guess is further away from the correct answer).

P (Guess)	$583.11 \times (1 + P)^{12}$	Higher / Lower (than target of 769.74)
0.04 (4%)	$583.11 \times (1 + 0.04)^{12}$ = 933.577897	Too High
0.02 (2%)	$583.11 \times (1 + 0.02)^{12}$ = 739.5244728	Too Low
0.03 (3%)	$583.11 \times (1 + 0.03)^{12}$ = 831.3754307	Too High
0.025 (2.5%)	$583.11 \times (1 + 0.025)^{12}$ = 784.2181223	Too High
0.024 (2.4%)	$583.11 \times (1 + 0.024)^{12}$ = 775.0861366	Too High
0.023 (2.3%)	$583.11 \times (1 + 0.023)^{12}$ = 766.0517234	Too Low
0.0235 (2.35%)	$583.11 \times (1 + 0.0235)^{12}$ = 770.5567929	Too High

We want to know the answer for P to 1 decimal place, since 2.4% is too high and 2.3% is too low, we know that the real answer for P is less than 2.4% and greater than 2.3% (i.e. that it is between 2.3% and 2.4%). This means that to 1 decimal place, the answer must either round to 2.3%, or to 2.4%, as shown in the diagram of the number line below.

If P is between 2.3% and 2.35%, then we round down to 2.3%, and if P is between 2.35% and 2.4%, then we round up to 2.4%. This is why, for our final guess, we trial the value of 2.35%. Since 2.35% is too high, our actual value of P is below this and hence rounds to 2.3%, to one decimal place. Therefore, our final answer for the actual average percentage that the shares grew by each year, to one decimal place, is 2.3%.

e) The value of the shares can go up or down in any given year, whereas the savings account guarantees to pay 2% interest. Just because a better rate was achieved in these twelve years, that does not mean that a better rate will always be achieved. Therefore, it is not correct, that he is always better off investing in shares, as opposed to investing in a savings account.